Vibration Analysis and Predictive Technologies in Reliability Engineering

by

Johnnie R. Ciulla Jr

Masters Degree of Science

Mechanical Engineering

2009

ACKNOWLEDGEMENT

I would first like to express my gratitude for my research supervisor, colleagues, peers and family whose immense and constant support has been a source of continuous guidance and inspiration.

DECLARATION

I Johnnie R. Ciulla Jr, declare that the following dissertation/thesis and its entire content has been an individual, unaided effort and has not been submitted or published before. Furthermore, it reflects my opinion and take on the topic and is does not represent the opinion of the University.

Signature:

Dated:

ABSTRACT

Trends towards long-span, lightweight floors in steel–concrete composite construction are resulting in structures possessing low natural frequencies, and potentially susceptible to vibration problems. The most common source of vibrations is caused by human activities on the floor, however, in some instances, mechanically induced vibrations from air conditioning plant, etc. may also be problematic. In this paper the serviceability assessment of floor vibration occasioned by walking activities is considered. Over 30 years ago concerns were raised regarding vibrations induced by walking on steel–concrete composite floors that satisfied traditional deflection criteria. In response to these concerns, design criteria based on a simple impulsive loading function from a person rising onto the balls of the feet, and suddenly dropping onto the heels has been used as a measure of floor acceptability. More recently, design procedures have been developed that more realistically consider the excitation of the floor from walking activities. In the first part of this paper a historical review of acceptability criteria for floor vibrations will be made. Following this, the current design methodologies for steel–concrete composite floors will be reviewed prior to the presentation of results from a series of in situ vibration tests on a wide variety of floor types. After comparing the test results with current guidance, design recommendations and suggestions for future research will be given.

CHAPTER 1: INTRODUCTION

The use of steel–concrete composite floor systems for multi-storey construction has increased dramatically over recent years. Such increase is largely due to the response of the construction industry to clients' demands for buildings that are fast to construct, have large uninterrupted floor areas and are capable of accommodating a high degree of servicing. Coupled with this are the requirements for reducing cladding costs and minimizing construction depths where building heights are restricted. In response to these demands, structurally efficient composite floor systems have been developed, which in some cases, are capable of incorporating services within the structural depth.

However, as a consequence of these developments, serviceability issues relating to the vibration response of the floor from walking activities are now frequently becoming the governing design criteria because: long slender spans can be readily achieved; the natural frequencies of the floor can be low; the construction is relatively light; and the level of damping is generally low. The recent highly publicized vibration serviceability problem with the Millennium Bridge in London1 has focused designer's attention to issues related to human acceptance of vibration.

In the past, many design guides and codes of practice have used the impulsive load due to a person rising onto the balls of the feet, and suddenly dropping onto the heels (known as a 'heel-drop' test) as a measure of acceptable floor performance. Conversely, in UK and Europe a minimum floor frequency has been used as a sole measure of acceptable performance5; this is based on the assumption that a sufficiently high natural frequency means that a floor is effectively 'tuned' out of the frequency range of the first harmonic component of the walking activity. However, many modern standards have now moved away from these traditional measures of vibration performance, and require that the designer make realistic estimates of the level of floor response that will be encountered in service by considering the walking excitation directly6, 7 and comparing this with human annoyance criteria.

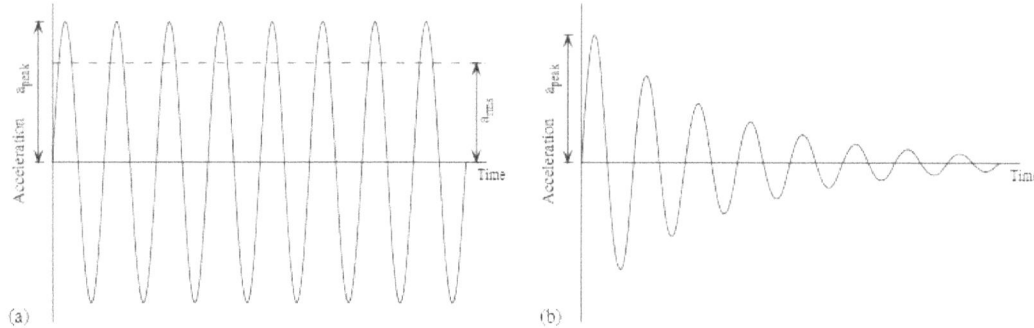

Fig. 1 Acceleration waveforms for sinusoidal motion: (a) continuous steady-state response; (b) transient response

CHAPTER 2: LITERATURE REVIEW

Continuous And Impulsive Vibrations

There are many possible ways in which the magnitude of the vibration response can be measured. For large-amplitude, low-frequency motion, it may be possible to observe the displacement between the maximum (i.e. peak) movement in one direction, and the peak movement in the opposite direction (i.e. the peak-to-peak displacement). In practice this distance can be difficult to measure and, for high-frequency motion, the vibration can be severe, even when the displacement is too small to be detected by the eye. The velocity, which is more directly related to the energy involved in the movement, may also be used to define the magnitude of the vibration. However, instrumentation for measuring acceleration is normally more convenient.

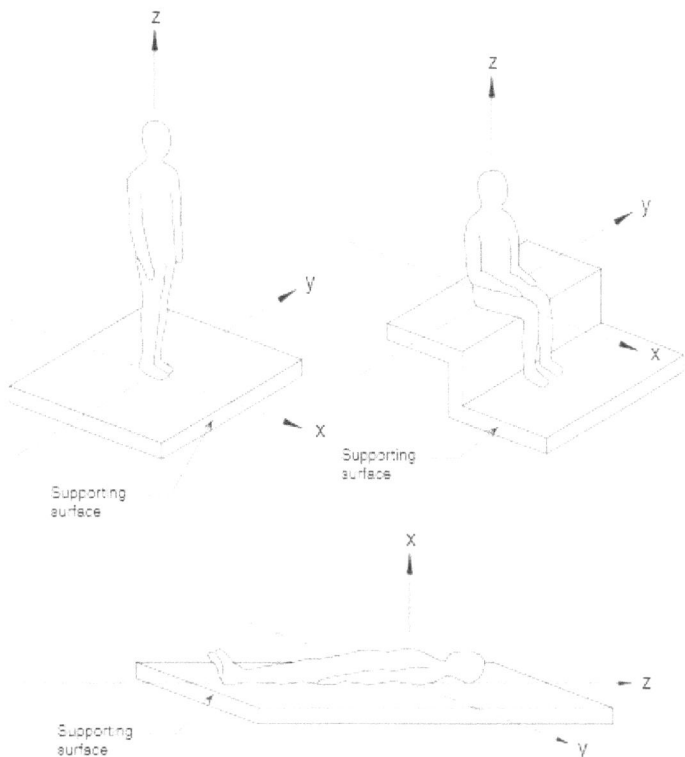

Fig. 2 Directions of basicentric coordinate systems for vibrations influencing humans

To ensure that the measure of the vibration is not influenced by one unrepresentative peak in the response, it is often preferred to express the severity of the vibrations in terms of an average measure. The measure in greatest use in current practice is the root-mean-square (rms) value (i.e. the square root of the average of the squared values), which may be expressed mathematically as follows:

where $a(t)$ is the acceleration time-history and T is the selected time period (in seconds). For continuous steady-state sinusoidal motion (Fig. 1a), the magnitude of the rms acceleration is the peak acceleration amplitude, 0.707 a(peak).

Extensive research on the human perception of vibration has been carried out on a wide variety of applications. This complex topic is dependent on a number of interrelated factors such as: the type of activity; the time of day when the activity is being undertaken; and the type of environment where the activity is taking place. Other vibration parameters such as the: direction; amplitude; frequency; source; damping; and the duration of the exposure are also important factors. The output from research on the human perception of vibration is often in the form of graphs indicating regions of acceptable or unacceptable vibration. The bounds given on these graphs depend on the direction of incidence to the human body, and generally use the reference coordinate system shown in Fig. 2 (the *z*-axis corresponds to the direction of the human spine). The most common graphs used to assess the acceptability of floor vibrations vary widely in their presentation by using different measures on each axis, and by considering damping in different ways.

So that direct comparisons can be made between the various recommendations, where appropriate, the graphs have been converted to a common format of rms acceleration plotted against frequency (Fig. 3). Each line shown on Fig. 3 represents a constant level of human reaction known as an isoperceptibility line, the area above a line corresponds to an unacceptable human reaction, while the area below represents acceptable levels of vibration. Unless noted otherwise, all of the graphs shown in Fig. 3 correspond to vertical vibrations for people in standing and sitting positions (i.e. z-axis vibrations, Fig. 2).

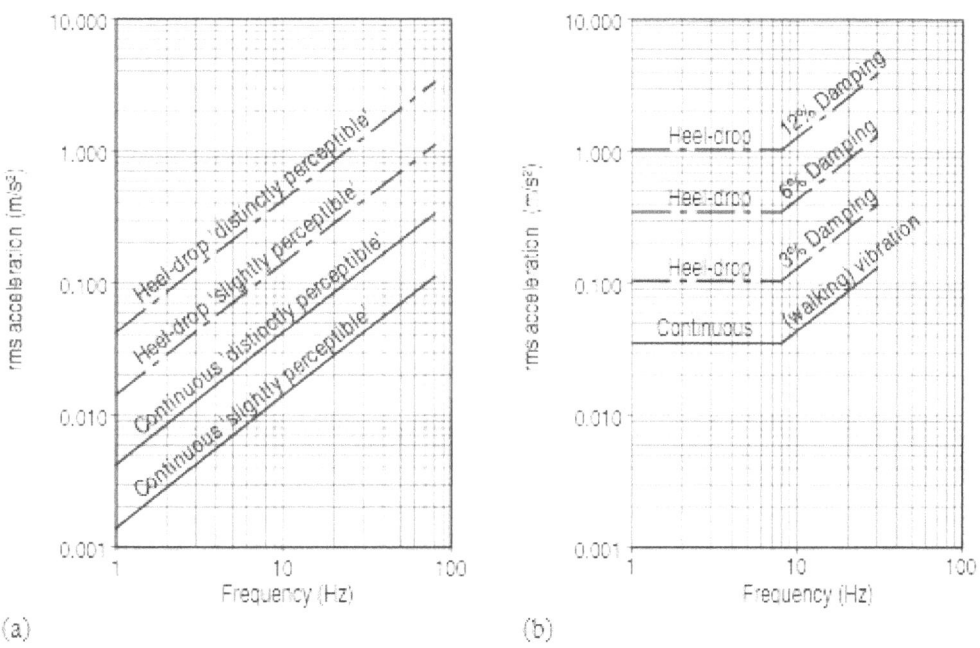

Figure 3. Human perception graphs in terms of rms acceleration plotted against frequency: (a) Modified Reiher and Meister9; (b) CAN3-S16.1-M892; (c) Murray12; (d) ANSI S3.29-198313, BS6472:199414, ISO 2631-2: 198215; (e) DIN 4150-2: 197516

One of the early studies on the perception of humans to whole-body vibrations was conducted in the 1930s by Reiher & Meister8. In this work, 10 subjects were subjected to continuous vertical and horizontal steady-state vibrations (i.e., of the form shown in Fig. 1a) in seated and reclining positions. In one of the first studies on the vibration of steel-framed floors Lenzen9 used the floor response to a single impact as a measure of acceptability. Since the vibration response from such a test was transient in nature (Fig. 1b), Lenzen modified Reiher & Meister's graph by increasing the acceleration for the isoperceptibility lines by factor of 10 to account for the reduced human sensitivity to transient vibration as opposed to steady-state vibrations, see Figure 3(a).

Based on research by Allen & Rainer10, annoyance criteria for floor vibrations were given in the 1974 and 1989 editions of the Canadian Standards Association (CSA) Standard for steel-framed buildings2 (Fig. 3b). The CSA scale requires careful interpretation, in that the assessment of a floor by the walking vibration lines, is based on the initial peak acceleration, frequency and damping of vibrations arising from a heel-drop test (which results in a response of the type shown in Fig. 1b). According to the CSA, the curves given in Fig. 3(b) are applicable for quiet occupancies in residential, school and office buildings; it is also noted that people are most sensitive to frequencies between 2 and 8 Hz, where the threshold of perception is approximately 0.5%g. Pernica & Allen11 later suggested that this threshold could be increased by a factor of 3 for 'active' environments such as shopping malls. Finally, based on field measurements of 91 steel–concrete floors, Murray12 proposed a series of curves based on the heel-drop test (Fig. 3c). He found a strong dependence on the damping within the floor for transient vibrations of this type, which show reasonable agreement with those adopted in the CSA guidance2.

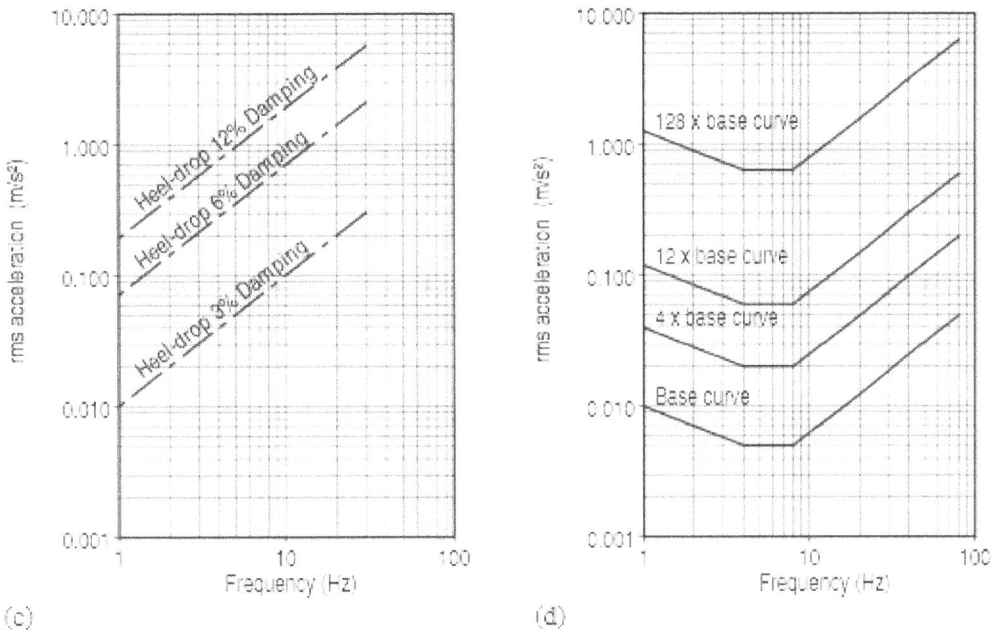

(c)   (d)

The American Standard ANSI S3.29-1983[13], British Standard BS 6472: 1992[14] and International Standard ISO 2631-2: 1982[15] are probably the most widely used codes of practice in modern design. However, rather than specifying specific levels of acceptable vibration, a 'base curve' is given, together with multiplying factors ranging from 1 to 128 for different environments (Table 1). The base curve, together with typical range of factored curves, is shown in Fig. 3(d). The German standard DIN 4150-2: 1975[16] uses an intensity perception parameter KB, which is established for each type of occupancy, usage, time of day and type of vibration (Table 1). After converting the vertical axis to rms acceleration, the isoperceptibility lines for a range of typical KB factors are presented in Fig. 3(e).

Place	Time	Multiplier for BS647214base curve (ANSI/ISO similar)		DIN 4150-2: 197516perceptibility parameter KB	
		Continuous vibration	Transient vibration	Continuous vibration	Transient vibration
Critical working areas	Day	1	1	0.1–0.6	4–12
	Night	1	1	0.1–0.4	0.15–0.4
Residential	Day	2–4	60–90	0.2	4
	Night	1.4	20	0.15	0.15
Office	Day	4	128	0.4	12
	Night	4	128	0.3	0.3
Industrial	Day	8	128	0.6	12
	Night	8	128	0.4	0.4

Table 1. Acceptability parameters as a function of the environment and the type of vibration

It should be noted, however, that the KB curves shown in Fig. 3(e) are appropriate for cases where the direction of the human occupants varies or is unknown, and therefore correspond to the worst-case curve for vibrations in the x-, y- and z-axis directions (Fig. 2). The KB curves are almost identical to those given in VDI 2057-2: 198717, which also considers direction dependant KX, KY and KZ values. In this case, the KZ curve corresponds exactly with the z-axis base curve given in ANSI S3.29-198313, BS 6472: 199214 and ISO 2631-2: 198215.

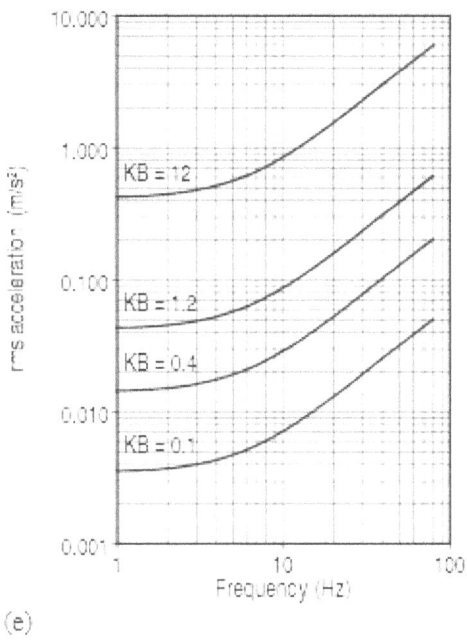

(e)

Intermittent Vibrations

All of these recommendations are based on continuous or impulsive vibrations. However, walking activities fall between these two categories into a third type termed intermittent. For intermittent vibrations a cumulative measure of the response has been found to be more reliable18. BS 647214 and ISO 2631-1: 199719 give guidance on this third category through vibration dose values (VDVs), where occasional short-duration exceedance of the continuous vibration levels given by Table 1 are tolerated if compensated for by periods of lower vibration. The general expression for calculating VDVs is shown below:

where VDV is the vibration dose value (in $m/s^{1.75}$), $a(t)$ is the acceleration time-history, which is 'weighted' according to a required dependence on vibration frequency19, 20 and T is the total period of the day during which vibration may occur (in seconds).

BS 6472 provides guidance for VDVs above which various degrees of adverse comment may be expected in residential buildings for both day- and night-time exposure. These values are presented in Table 2.

Table 2. BS6472[14] vibration dose values ($m/s^{1.75}$) above which various degrees of adverse comment may be expected in residential buildings			
Place	**Low probability of adverse comment**	**Adverse comment possible**	**Adverse comment probable**
Residential buildings 16-hour day	0.2–0.4	0.4–0.8	0.8–1.6
Residential buildings 8-hour night	0.13	0.26	0.51

By the very nature of eq. (2) the vibration needs to be recorded over the complete exposure period (i.e. from Table 2, a 16-h day or an 8-h night). However, in practice, it will often not be possible to undertake measurements over such long periods, and assessments may need to be made on floors based purely on calculations. This is recognised in BS 6472, which gives the following equation to calculate estimated vibration dose values (eVDVs):

$$VDV = \left(\int_0^T a(t)^4 dt \right)^{1/4}$$

where eVDV is the estimated vibration dose value (in $m/s^{1.75}$), *a(rms)* is the rms acceleration (in m/s^2) and t is the total duration of the vibration exposure (in seconds).

Unfortunately, although used in most of the code of practice perception scales for continuous vibrations[13–15], the rms acceleration is not an ideal measurement when considering walking vibrations. This is due to the fact that the value of the rms acceleration is strongly affected by the length of the selected time period (eq. 1). For example, for a very short time period, the rms value would tend towards a value of *a(peak)*/√2; whereas, if a long time period were selected, a much lower rms value would result. As a consequence of this, eq. (3) is only useful in estimating VDVs when the vibration is 'well-behaved'18 (i.e. the vibration is not intermittent, there are no shocks or other time-varying conditions), and therefore is of little use when considering vibrations due to walking activities.

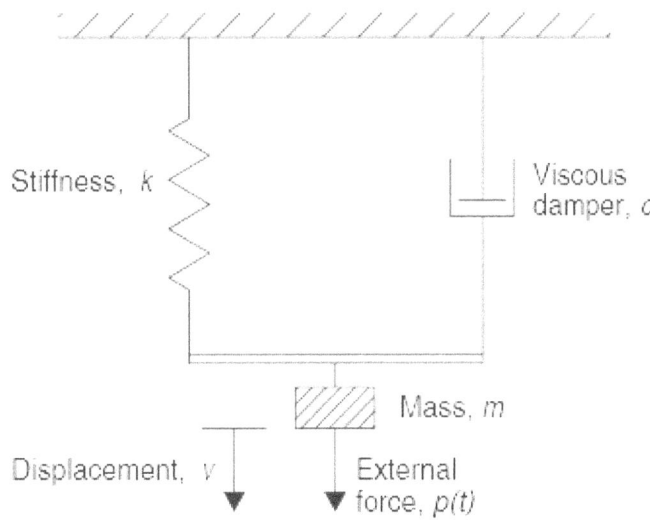

Fig. 4 Simple single-degree-of-freedom spring mass

In an attempt to provide guidance for evaluating VDVs based on calculations, five steel-framed floors subjected to walking activities have recently been monitored by Ellis21. Based on this study, the following equation was recommended for calculating a representative vibration dose value rVDV:

where a(peak) is the peak acceleration, which is 'weighted' according to a required dependence on vibration frequency19, 20 and t is the total duration of the vibration exposure (in seconds).

For example, based on eq. (4), ten occurrences, each of 16 s duration with vibration levels equivalent to a base curve multiplier of 25 would be tolerated in an office each day (Table 2), provided that vibrations at other times were low.

Although it has been acknowledged18 that the VDV approach is a more reliable measure of vibration response, it is not widely used in the design of floors, as walking corridors are often not known in the early design stage, and it is normally required that there is some flexibility in the floor layout. Furthermore it has been claimed that, although the VDV approach has been established for a variety of vibration sources, it may not be entirely appropriate for floors subjected to walking activities. Also, on very sensitive floors (such as operating theatres and precision laboratories), some guidelines do not permit making allowance for the intermittency of events; resulting in the multipliers shown in Table 1, or other more stringent values having to be used. As a consequence of this, most modern design guides conservatively assume that walking activities produce continuous vibrations on steel-concrete composite floors.

CHAPTER 3: METHODOLOGY

Review of existing design methodologies

The two design guides that have particular relevance to the design of steel–concrete

composite floor systems are the Steel Construction Institute (SCI) Publication 07625 and the

American Institute of Steel Construction/Canadian Institute of Steel Construction (AISC/CISC)

Design Guide 1126. Both guides are similar in that they idealize the floor as a simple single-

degree-of-freedom (SDOF) spring mass (Fig. 4), which is excited by the force from a single

person engaged in a walking activity (the heel-drop excitation previously used in North America

is no longer considered in the AISC/CISC design guide).

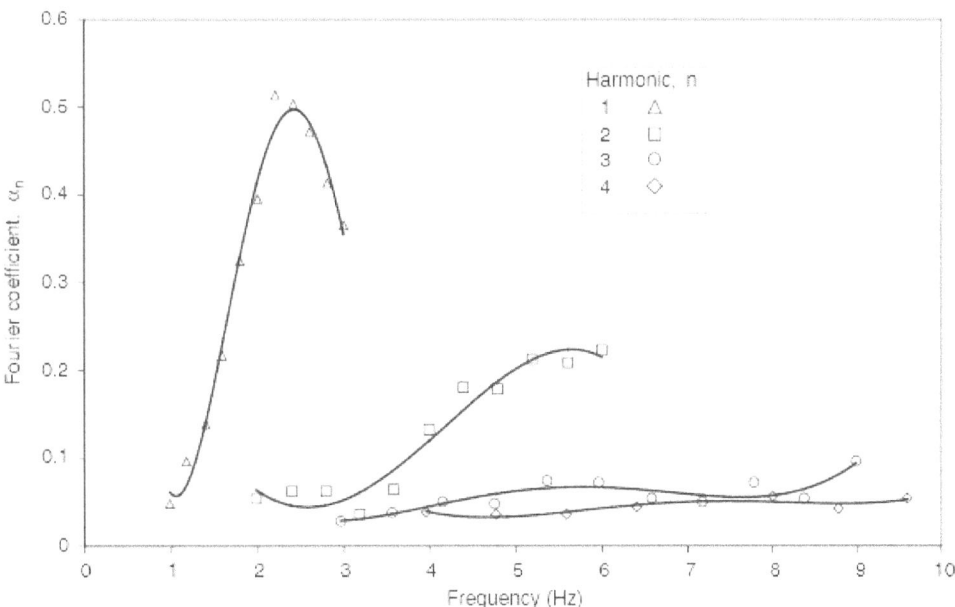

Acceptable Floor Performance

Multipliers to the base curve shown in Fig. 3(d) are defined as 'response factors' R in the

SCI design guide. For office accommodation, it is recommended that the response factor should

not exceed 4 for a 'special office', which is described as a suitable environment for technical tasks requiring prolonged special concentration (this value is also recommended for large public circulation areas such as pedestrian malls, extensive lobbies, banking halls, etc.). A response factor of 8 is given for a 'general office' classification, which is described as an environment that is suitable for normal office activities. While a response factor of 12 is suggested for a 'busy' office classification, which is described as an environment accessible to a large number of persons, with both visual and audible distractions concurrent with any vibration.

Although based on the base curve shown in Fig. 3(d), the AISC/CISC guide considers amplitudes of peak acceleration as a measurement of acceptable floor performance. For office, residential and church buildings it is recommended in this guide that the peak acceleration should not exceed 0.5%g (this is identical to the value given by the continuous walking line shown in Fig. 3(b) for 'quiet occupancies'); based on the base curve shown in Fig. 3(d), this is equivalent to a multiplying factor of 7. For shopping malls, a higher peak acceleration value is permitted with a value of 1.5%g, which is equivalent to a base curve multiplier of 21 (this value is identical to the limit recommended by Pernica & Allen[11]).

$$a_{peak} = \frac{x_n P_0}{M} \frac{1}{2\zeta}$$

Floor Response

Although floors will have many modes of vibration, in practice only a few modes (usually at the lowest frequency) will significantly contribute to the overall response. As a consequence of this, both the SCI and the AISC/CISC guide assume that the largest acceleration levels are produced when the walking activity excites the first mode of vibration (sometimes referred to as the 'fundamental mode'), and therefore only consider this mode in the estimation of the floor response. Two types of excitation are normally assumed to occur within floors that are subjected to walking activities—resonant excitation; and impulsive excitation. A fuller description of these two types of excitation is given below.

Resonant excitation

A load that varies sinusoidally with time at a constant frequency is known as a harmonic load. When such a force of amplitude p_0 and frequency f is applied to the simple structure shown in Fig. 4, the structure will be caused to vibrate. After some time the motion of the structure will reach a steady-state (Fig. 1a); the ratio of the resultant response amplitude to the static displacement, which would be produced by the force p_0, is called the dynamic magnification factor D, and is given the following identity:

where β is the ratio of the applied load frequency to the natural frequency of the system and ζ is the damping ratio.

When the dynamic force is applied at a frequency close to the natural frequency of a structure, which is lightly damped (as is found in most practical composite floor systems), the peak steady-state response will occur. The condition when the frequency of the applied load equals the natural frequency of the structure is called *resonance*. In these circumstances, very large dynamic magnification factors are possible and, for undamped systems (i.e. $\zeta=0$) the steady-state response tends towards infinity. A more general result may be obtained from eq. (5) which shows that for resonance ($\beta=1$) the dynamic magnification factor is inversely proportional to the damping ratio, and:

A repeated force, such as walking, can be represented by a combination of sinusoidal forces, whose frequencies are multiples (or harmonics) of the basic frequency of the force repetition (e.g., the pace frequency). For many practical purposes, the load-time function can be expressed by the following Fourier sum:

where P_0 is the static force exerted by the person, α_n is the Fourier coefficient of the nth harmonic, f_p is the activity frequency, t is the time (s), φ_n is the phase angle of the nth harmonic relative to the first harmonic and n is number of the nth harmonic.

In one of the first investigations that considered the loading produced by a variety of human activities, it was reported by Rainer et al.[27] that only the first four harmonics of the pace frequency comprise the main dynamic components of walking forces. From this study, the Fourier coefficients α_n shown in Fig. 5, for an individual walking at between 1.0 and 3.0 Hz, were established; subsequent investigations have produced similar values[28–32]. A summary of the average Fourier coefficients produced by these investigations is presented in Table 3.

Table 3. Average values of Fourier coefficients for walking activities					
Reference	Activity rate (Hz)	Fourier coefficient			
		First harmonic α_1	Second harmonic α_2	Third harmonic α_3	Fourth harmonic α_4
Rainer et al.27	1.0–3.0	0.33	0.14	0.05	0.05
Ellis28	1.7–2.4	0.46	0.09	0.07	0.08
Alves et al.29	1.6–2.0	0.34	0.11	0.11	–
Kerr30	1.0–2.8	0.32	0.07	0.05	0.05
CEB Bulletin d'Information31	1.6–2.4	0.40	0.10	0.10	–

Table 3. Average values of Fourier coefficients for walking activities					
		Fourier coefficient			
Reference	Activity rate (Hz)	First harmonic α_1	Second harmonic α_2	Third harmonic α_3	Fourth harmonic α_4
ISO/DIS 10137: 199032	1.7–2.3	0.40	0.20	0.06	–

As mentioned already, both the SCI and the AISC/CISC guides assume that the largest acceleration levels are produced when the walking activity excites the first mode of vibration. Other simplifications that are made by both the SCI and AISC/CISC guides for the case when the walking activity causes resonance to the floor are as follows:

- Since the largest accelerations are generated when the floor's fundamental frequency is an integer multiple (harmonic) of the pacing frequency, and as the pace frequency falls within well-defined boundaries (typically 1.6–2.4 Hz31, 33, an appropriate walking frequency can be selected to cause resonance to the floor.

- The reason for the largest accelerations being generated at an integer multiple of the pacing frequency is that resonance is encountered and, in this situation, the response at one frequency is dominant. Therefore, it is necessary to consider the response at that frequency only, and just one of the Fourier terms of the loading is needed (generally, the lowest multiple, or harmonic, of the pacing frequency will give the largest acceleration, Table 3).

By idealizing the floor as the simple SDOF system shown in Fig. 4, the SCI and AISC/CISC guides give the following equation for the positive peak acceleration at steady state, in different forms:

where α_n is the Fourier coefficient of the nth harmonic, P_0 is the static force exerted by an 'average person' (normally taken as 76 kg × 9.81=745.6 N), M is the modal mass and ζ is the damping ratio.

For the SCI guide, the Fourier coefficients for the second and third harmonic components are based on the work by Rainer *et al.*27 and are given in terms of a trilinear equation which expresses the average amplitude of the Fourier coefficient as a function of the fundamental floor frequency. To avoid a step function for floors with a fundamental frequency between 4.0 and 4.8 Hz (i.e. the transition point between the second and third harmonic components of the walking activity), the magnitude of the Fourier coefficient is reduced linearly in this guide. The SCI guide also uses a design walking force that is much smaller than that exerted by an 'average person' with a magnitude of 240 N.

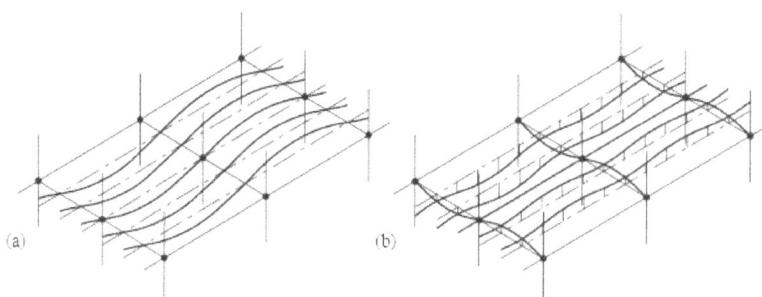

Fig. 6 Typical fundamental mode shapes for composite floor systems: (a) governed by secondary (joist) beam flexibility; (b) governed by primary (girder) beam flexibility

The AISC/CISC guide adopts a similar procedure to the SCI guide by expressing the amplitude of the Fourier coefficients as a function of the fundamental floor frequency. However, in this case, an exponential equation is specified (which is based on a pace frequency of between

1.6 and 2.2 Hz). For this guide, the average value of the Fourier coefficients for the first, second, third and fourth harmonic components are 0.5, 0.2, 0.1 and 0.05, respectively (Table 3). In a similar way as the SCI guide, the force exerted by an average person is multiplied by a reduction factor to give a design walking force for floors of 290 N.

Both the SCI and AISC/CISC take different approaches in establishing the modal mass term shown in eq. (8). For the SCI guide, the design rules were developed from a parametric study of simply supported orthotropic plates34, and depend on whether the fundamental frequency is dominated by motion of the secondary (joist) or primary (girder) beams. While, for the AISC/CISC guide, the modal mass is calculated more directly by considering the stiffness between: the slab and the joists (secondary beams); the joists and the girders (primary beams); and how these elements are connected to one another.

Finally, for the damping term given in eq. (8), both the SCI and AISC/CISC recommend similar values. For exceptionally bare steel–concrete composite floors, as might be encountered just after construction, damping values of 1.5 and 1.0% are suggested for the SCI and AISC/CISC guides respectively. For normal, open-plan, well-furnished floors a value of 3.0% is recommended. While, for cases when full-height partitions are present between floors, damping values of 4.5 and 5.0% are recommended by the SCI and AISC/CISC guides, respectively.

CHAPTER 4: RESULTS AND ANALYSIS

Impulsive excitation

Because most of the excitation energy is concentrated within the lower harmonic components of the walking activity (Fig. 5), for floors that posses a sufficiently 'high' frequency, such that the first four harmonic components of the walking force do not cause resonance, the response is dominated by a train of impulses, which correspond to the heel impacts. The basic effect of these impulses is that they set the mass of the floor in motion, which vibrates at its natural frequency and decays rapidly as energy is dispersed over the floor as a whole. As a consequence, successive peaks and decays typify the overall dynamic response of a floor of this type (the idealized response from a single 'footstep' is shown Fig. 1b). For the SCI guide, the transition point between a response of this type and the case when resonance occurs is defined as floors that possess a fundamental frequency greater or equal to 7.0 Hz, whereas, for the AISC/CISC guide, this transition point is higher, and is defined as floors that have a fundamental

frequency 9.0–10.0 Hz.

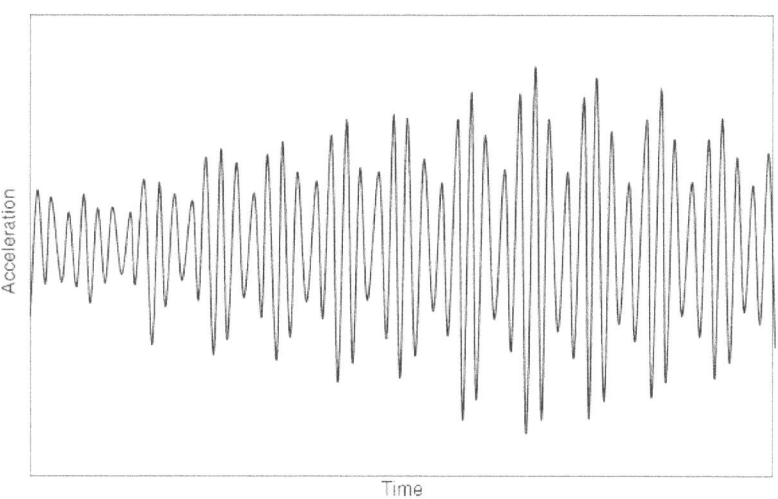

Fig. 7 Resonant build-up of a floor from an individual with a pace frequency of 1.6 Hz (the increasing peaks at every fourth cycle indicate that the floor is being excited by the fourth harmonic component of the pace frequency)

In order to determine the response for floors of this type, the SCI guide again considers the simple SDOF system shown in Fig. 4, but this time considers the following equation for calculating the peak acceleration due to an impulsive force:

where I is the impulsive force (N) and M is the modal mass.

For this type of excitation, the SCI guide assumes that the force exerted by a heel-strike corresponds to a value of 3–4 N s. For the determination of the modal mass, the SCI guide assumes a very conservative value, which corresponds to an area equivalent to the beam span multiplied by its spacing.

No specific guidance is given within the AISC/CISC guide for impulsive excitation, because this case is taken into account35 through the variables used in eq. (8), together with a floor stiffness requirement of 1.0×10^6 N/m under a concentrated load of 1000 N.

Natural Frequency

For free elastic vibration of a beam, or uniform section, the natural frequency is given by:

where EI isdynamic flexural rigidity of the member (N m^2), m is the effective mass (kg/m), L is the span of the member (m), and k_n is a constant representing the beam support and/or loading conditions.

Some standard values of k_n for the first mode of vibration of elements with different boundary conditions are as follows36: 4

Table 4.	
pinned/pinned ('simply supported')	π^2
fixed/pinned (propped cantilever)	15.4
fixed both ends (encastré)	22.4
fixed/free (cantilever)	3.52

A convenient method of determining the natural frequency of a beam f, is presented within the SCI guide25, by first finding the maximum deflection δ (in millimetres) caused by the weight of a mass m. For a simply supported element subjected to a uniformly distributed load ($k_1=\pi^2$), this is of course:

where g is the acceleration due to gravity (i.e. 9.81 m/s^2).

Rearranging eq. (11), and substituting the value of m and k_1 into eq. (10), gives:

For beams with different loading types and/or boundary conditions, similar results can be found with the numerator in eq. (12) varying between 16 and 20. However, for practical design, a value of 18 will normally produce results of sufficient accuracy.

According to the SCI guide, the maximum deflection δ given in eq. (12) should be based on the gross second moment of area of the composite beam, with a load corresponding to the self-weight, and other permanent loads, plus a proportion of the imposed load that may be considered as permanent (taken as 10% for offices). A similar expression is given within the

AISC/CISC design guide. However, in this case, a constant imposed load of 0.5 kN/m^2 is suggested for offices.

The above concept can be extended to enable an estimate of the fundamental (first mode) frequency f_0 of a complete composite floor system. This is achieved by using engineering judgement on the likely mode shape and the support conditions this will impose on the individual structural components. For example, the SCI guide suggests that on a simple floor comprising a slab continuous over a number of secondary (joist) beams, which in turn, are supported by stiff primary (girder) beams (Fig. 6), there are two possible mode shapes that may be sensibly considered:

- Secondary (joist) beam mode: the primary beams form nodal lines (i.e. they have zero deflection) about which, the secondary beams vibrate as simply supported members (Fig. 6a). In this case, the slab flexibility is affected by the approximately equal deflections of the supports and, as a result of this, the slab frequency is assessed on the basis that fixed-ended boundary conditions exist.

- Primary (girder) beam mode: the primary beams vibrate about the columns as simply supported members (Fig. 6b). By similar reasoning, because the equal deflections at their supports, the secondary beams (as well as the slab) are assessed on the basis that fixed-ended boundary conditions exist.

The frequency of the whole floor system can be calculated for each mode shape, by summing the deflection calculated from each of the above components, and placing this value within eq. (12). The lowest frequency value determined by consideration of these two cases defines the fundamental frequency of the floor f_0 (and its corresponding mode shape). Alternatively, it can sometimes be convenient to use the component frequencies directly, to evaluate the fundamental frequency of the floor by Dunkerly's approximation shown in eq. (13); both methods give identical results.

where f_{c1}, f_{c2} and f_{c3} are the component frequencies (Hz) of the composite slab, secondary beams and primary beams respectively, with their appropriate boundary conditions, as defined above.

The AISC/CISC design guide adopts a similar procedure for estimating the fundamental frequency of a composite floor. However, in this case, three modes of vibration are considered: a 'beam mode' (similar to Fig. 6a); a 'girder mode'; and a 'combined mode' (in a similar way to Fig. 6(b), a combination of deflections from both the beams and the girders). Also, the AISC/CISC guide does not consider the contribution from the slab deflection, and assumes that the beams are simply supported.

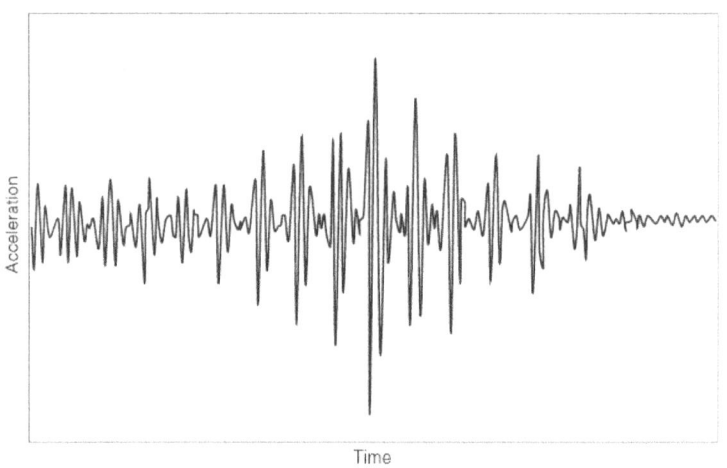

Fig. 8 Transient response from a person engaged in walking activity at 2.5 Hz.

Vibration tests on steel-concrete composite floors

Dynamic testing has recently been carried out on a variety of composite floor structures37. In an effort to cover as wide a range of floor types as possible, a total of 18 floors have been tested; brief descriptions of the tested structures are given in Table 4. The choice of floor was largely governed by availability, as many owners of buildings were reluctant to grant access for fear of causing alarm to occupants. Therefore, in the majority of cases, the floors tested were in their unoccupied state just after construction.

Floor	Project	Structural form	Finishes	Test panel dimension NS × EW (m)
		Table 4. Details of buildings tested		
1	Test building at BRE Cardington38	Composite floor, 9 m span, 305×165UB40 secondary beams EW spaced at 3 m centres NS. 9 m span 610×229UB101 NS primary beams	None	21 × 14
2	Paris Office Area 1	Long-span composite floor, 13.75 m span, 459 mm deep cellular beams NS spaced at 2.7 m centres EW	Services	13.75 × 14.15
3	Paris Office Area 1	As above	False floor, services	13.75 × 14.15
4	Paris Office Area 2	Long-span composite floor, 15.68 m span, 459 mm deep cellular beams EW spaced at 2.7 m centres NS, 7.5 m span, 459 mm deep primary cellular beams	Services	23.19–13.5 (tapered) × 5.5–22.95 (tapered)
5	Paris Office Area	As above	False floor,	23.19–13.5

Floor	Project	Structural form	Finishes	Test panel dimension NS × EW (m)
	2		services	(tapered) × 5.5–22.95 (tapered)
6	Paris Office Area 3	Long-span composite floor, 16.65 m span, 459 mm deep cellular beams NS spaced at 2.7 m centres EW	False floor, services	16.6–13.2 (tapered)×32.4
7	Cambridge Laboratory	Slimdek® floor, 6 m span, 280ASB136 NS spaced at 6.6 m, centres EW	Services	20.11 × 26.4
8	SCI Headquarters39	Composite floor, 6 m span, 305×127UB42 secondary beams EW spaced at 2.5 m centres NS; 7.45 m span, 686 × 152UB60 castellated primary beams NS	False floor, services, furniture, partitions	14.9 × 24
9	London Office 1	Long-span composite floor, 15.31 m span, 742 mm deep cellular beams EW spaced at 3 and 1.5 m centres NS	False floor, ceiling, services	42.17×15.31
10	London Office 2	Long-span composite floor, 15 m	None	15–11.35

Table 4. Details of buildings tested

Table 4. Details of buildings tested				
Floor	Project	Structural form	Finishes	Test panel dimension NS × EW (m)
		span, 664 mm deep cellular beams NS spaced at 3 m centres EW		(tapered) ×17.23
11	London Office 3	Long-span composite floor, 10.5 m span, 400 mm deep cellular secondary beams EW spaced at 2.6 m centres NS, 10.5 m span, 508 mm deep cellular primary beams NS	None	18 × 31.5
12	London Office 3	As above	False floor, ceiling, services	18 × 31.5
13	London Office 4	Long-span composite floor, 13.5 m span, 500 mm deep cellular secondary beams NS spaced at 3 m centres EW	None	13.5 × 45
14	Lloyds Corporation 1958 Building40	Long-span composite floor, 5.6 m span, 152×152UC30 secondary beams EW spaced at 2.4 m centres NS; 16.14 m span 546×406UC340	None	16.14 × 39.27

Floor	Project	Structural form	Finishes	Test panel dimension NS × EW (m)
		Castellated primary beams NS		
15	Lloyds Corporation 1958 Building40	As above	False floor, services	16.14 × 39.27
16	University of Wales41	Composite floor, 10.8 m span, 457 × 152UB52 secondary beams NS spaced at 2.6 m centres EW; 7.8 m span 610 × 229UB140 primary beams EW	None	18.3 × 20.8
17	University of Wales41	As above	False floor, services, office furniture, partitions	18.3 × 20.8
18	London Office 5	Slimflor® floor, 9 m span, 305×305UC97 NS spaced at 7.5 m centres EW	False floor	16.5 × 31.5

Table 4. Details of buildings tested

An explanation of the test procedure, and the particular dynamic properties that were gained, is discussed below.

Impact Tests

Impact tests, as the name implies, simply consist of disturbing the structure from its quiescent condition, through a single impulse, and monitoring the resulting response with an accelerometer. These tests are quick and simple to conduct, and are usually adequate for obtaining the *in situ* vibration properties of relatively simple structures. Typically, two types of excitation may be used in impact tests:

Heel-drop excitation

Heel-drop tests comprise of a single person rising on the balls of the feet, and suddenly dropping onto the heels: thus providing an impact to the floor, which can be monitored (Fig. 1b). As discussed in the earlier section on human perception, in the past, this simple loading function has been used within some design guides to assess the acceptability of floors[2–4]. However, a long-standing problem with this approach has been that the input force is not measured, and may vary from test to test. Some researchers have recently remedied this by developing an instrumented heel-drop test[42], where the heel-drop is executed on top of a slim, purpose-built load cell.

Instrumented hammer excitation

This method and associated signal processing techniques are described in detail elsewhere[43]. In essence, the method consists of striking the structure with a soft-tipped hammer, instrumented with a force transducer, and measuring the response of the floor using an accelerometer. The natural frequencies and damping values are then deduced from the transfer function between the two signals. By moving the hammer or the accelerometer to a predefined

set of grid points on the floor, the corresponding mode shapes may also be obtained *in situ*.

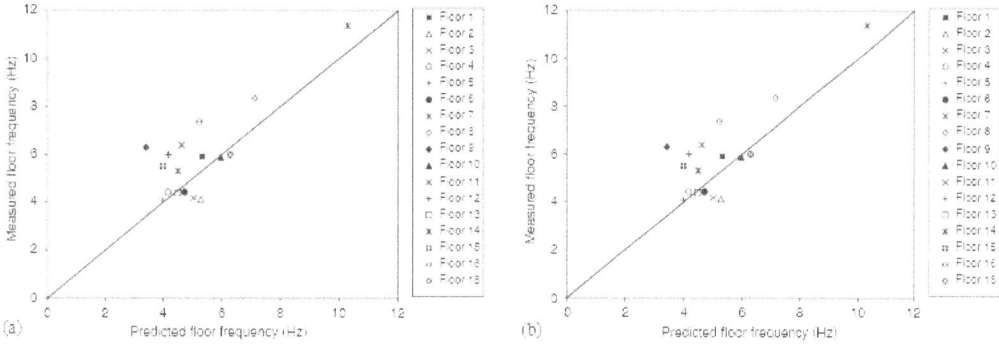

Fig. 9 Comparison of measured floor frequencies with predicted floor frequencies according to: (a) SCI publication 076[25]; (b) AISC/CISC Design Guide 11[26]

Forced Vibration Tests

Impulsive tests sometimes have difficulty distinguishing between closely spaced natural

frequencies. The existence of several very close modes of vibration is quite common in floors,

which often have several bays of similar stiffness in each span direction. In such cases, better

quality data may be obtained by providing a continuous forcing input to the floor. A number of

devices are currently available to do this; the simplest is the rotating mass exciter, in which two

masses rotate at the same speed, but in opposite directions, so that the horizontal components of

their inertia forces cancel, leaving only a sinusoidally varying vertical force40. More

sophisticated systems also exist, in which inertial masses are shaken by electromagnetic or

hydraulic forces.

Table 5 summarizes the measured fundamental (first-mode) frequency, the corresponding

modal damping ratio and stiffness for the test floors. The wide scatter of the natural frequencies

reflects the range of structural types tested.

Table 5. Measured fundamental frequencies, damping ratios and modal stiffness for the floors tested			
Floor	**Fundamental frequency (Hz)**	**Damping ratio (%)**	**Modal stiffness (N/m)**
• † Dubious result • ‡ Results affected by the presence of full-height blockwork partitions			
1	5.92	1.04	2.23×10^7
2	4.13	1.28	3.80×10^6
3	4.19	1.55	4.40×10^6
4	4.44	1.53	1.98×10^6
5	4.09	1.49	5.09×10^6
6	4.44	1.40	9.12×10^5
7	11.38	2.93	9.76×10^7
8	8.35	4.68	1.16×10^7
9	6.30	†	–
10	5.88	3.4	–
11	6.40	4.6	–
12	6.00	3.4	–
13	4.40	†	–

Table 5. Measured fundamental frequencies, damping ratios and modal stiffness for the floors tested			
Floor	Fundamental frequency (Hz)	Damping ratio (%)	Modal stiffness (N/m)
14	5.32	0.87	7.21×10^6
15	5.52	0.91	–
16	7.38	2.50	3.82×10^7
17	9.65‡	2.85	3.40×10^7‡
18	6.00		

Walking Tests

The aim of walking tests is to ascertain the worst (design) case for the response of floors in service. Walking frequencies are chosen so that, for cases when it is expected that the floor will exhibit a resonant response, their harmonics coincide with the natural frequencies of the floor. For example, for a floor with a fundamental frequency of 4.0 Hz it would be necessary to perform walking tests at 2.0 Hz (i.e. 2×2.0=4.0 Hz). The pace frequency can be controlled by walking in time to a beat generated by a portable computer or a metronome.

The choice of walking path is also important, as the greatest response will normally be achieved by planning a walking path that crosses the point on the floor where the mode shape takes its maximum value (i.e. the antinode). As a consequence of this, the walking path may vary depending upon which mode of vibration the walking test is attempting to excite.

A typical example of resonant build-up of a floor when subjected to a walking test is shown in Fig. 7. For cases when a floor possesses a sufficiently high frequency, such that the first four harmonic components of the walking force do not cause resonance, the characteristic transient response from a walking test is shown in Fig. 8.

As discussed earlier, the rms acceleration is not an ideal measurement when considering walking vibrations because the value is strongly affected by the length of the selected time period (eq. 1). In the UK, there is little information contained within the current codes of practice on a standard period that should be considered when calculating the rms acceleration for intermittent vibrations. As a consequence of this, researchers have often considered the 'significant' portion of the acceleration–time history from walking tests, which has caused the results of such tests to vary widely. A study in Sweden, by Eriksson[45], suggested that a period of 10 s is appropriate; however, this may be impractical for small floor areas, where only a few paces may physically be undertaken. ISO 2631-1: 1997[19], VDI 2057-1: 2002[46] and DIN 4150-2: 1999[16] suggest a 'running root-mean-square' value, where portions of the acceleration time history are considered using an integration time for running averaging, τ. DIN 4150-2: 1999[16] recommends a 'fast' rms value using $\tau=0.125\ s$. Although VDI 2057-1: 2002[46] also recommends an identical 'fast' integration constant, a 'slow' rms value corresponding to $\tau=1$ second is permitted (an integration time constant of $\tau=1s$ is also recommended in ISO 2631-1: 1997[19]).

When considering these recommended periods for walking activities, although the 0.125 s integration constant may be acceptable for cases when resonance occurs (the motion is generally repeatable over short periods), it may not be entirely appropriate for floors that exhibit a transient response, where the period between each footfall can be nearly five times longer than this value. For this reason, a period of 1 s has been used to obtain the rms values shown in Table

6. This period was sufficient to capture at least two footfalls in the acceleration time history for the floors that exhibited a transient response (Floors 7 and 8).

Table 6. Summary of measured floor responses from walking tests			
Floor	Measured peak acceleration (m/s^2)	rms acceleration(m/s^2)	Equivalent multiplying factor based on BS6472 z-axis base curve
1	0.076	0.054	10.7
2	0.041	0.029	5.8
3	0.105	0.013	2.6
4	0.045	0.031	6.3
5	0.039	0.028	5.6
6	0.059	0.042	8.3
7	0.059	0.026	3.9
8	0.133	0.020	4.1

Comparison of test results with current design guidance

In this section, the test results presented in Table 5 and Table 6 are compared with the predictions from the current SCI and AISC/CISC design guides. In interpreting the performance of the respective methodology, the ratio of the test value to that of the predicted value is used to give a dimensionless term, hereafter referred to as the model factor.

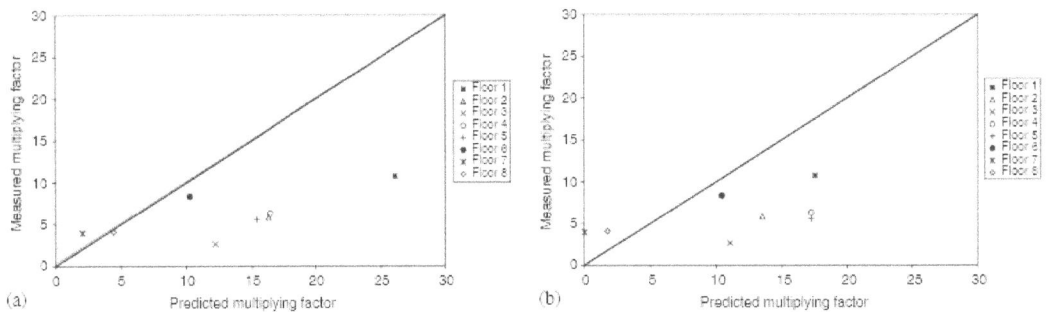

Fig. 10 Comparison of measured floor response with predicted floor response according to: (a) SCI publication 076[25]; (b) AISC/CISC Design Guide 11[26]

Fundamental Floor Frequency

As described earlier, two mode shapes exist which may be sensibly considered in design—a secondary (joist) beam mode; and a primary (girder) beam mode (Fig. 6). The lowest frequency value determined by consideration of these two mode shapes is the first mode, or fundamental frequency, of the floor structure. In assessing the level of response, it is presently assumed by both the SCI and AISC/CISC design guidance that the floor only has this mode of vibration25, 26.

From Table 5, the lowest frequency measured on Floors 1–16 and 18 is plotted in Fig. 9 against the fundamental frequency calculated in strict accordance with the current SCI and AISC/CISC guides, using the appropriate level of load present on the floor at the time of testing (Floor 17 is not included as the results are affected by full-height partitions).

As can be seen from Fig. 9, for the 17 floors considered, the design methodology given in both the SCI and AISC/CISC guides produce similar levels of correlation with the measured values. The average model factor for the SCI predictions is 1.01, with a corresponding coefficient of variation (COV) of 21%. For the AISC/CISC predictions, a slightly more conservative average model factor of 1.15 is achieved, but with a similar COV to the SCI guide of 23%. A similar conclusion has also been made from a recent study that considered a greater number of floors48. However, for floors that consisted of composite trusses (sometimes referred to as 'open web joists' in North America), it was reported[48] that the AISC/CISC approach produced more accurate predictions.

Damping

As described earlier, the SCI and the AISC/CISC guide[25, 26] suggest three damping values, which may be considered when assessing the response of a floor. These values are dependant on the level of non-structural elements on the floor, and the presence of partitions.

From a consideration of the measured values for the completely bare and semi-bare floors shown in Table 5 (where false flooring, ceiling and services were present; but no office furniture), it was found[37] that the level of damping was remarkably similar. Furthermore, the statistical properties for these two floor types were virtually identical. As a result of this, it has recently been concluded[37] that the presence of false flooring, ceiling and services does not contribute significantly to the damping characteristics of the floor. This finding also confirms observations made by others[40]. A summary of the measured damping values obtained from the floors described in the preceding section, are presented in Table 7.

Table 7. Summary of measured damping values for 18 floors tested		
Floor finishes	**Range of spans for damping values given (m)**	**Average damping ratio ζ (% of critical)**
Bare/false floor, ceiling and services	6–17	2.29
False floor, services and office furniture.	10.8	2.85

Table 7. Summary of measured damping values for 18 floors tested		
Floor finishes	**Range of spans for damping values given (m)**	**Average damping ratio ζ (% of critical)**
False floor, services, furniture and partitions	7.45	4.68

As can be seen from Table 7, for 'bare' floors the average measured damping value of 2.29% is significantly higher than that recommended in either the SCI or AISC/CISC guide (1.5 and 1.0%, respectively). However, because the scatter of the experimental results is vast (with a COV of 51%), it has been recommended[37] that an appropriate design value for bare floors is $\zeta=1.1\%$, which corresponds to the mean value of the damping minus one standard deviation: representing a 65% probability of exceedance. As a result of this, the design values given within the SCI AISC/CISC guides appear to be appropriate for this floor case.

Unfortunately, since the majority of floors tested were in their unoccupied state, damping values for floors with normal finishes, and floors that were heavily partitioned, were obtained only from two floors (Floors 17 and 8). As can be seen from Table 7 the experimentally determined damping value of $\zeta = 2.85\%$ for a normal furnished floor (Floor 17) compares well with the design value currently recommended by both the SCI and AISC/CISC guide ($\zeta=3.0\%$). Whereas, for a heavily partitioned floor (Floor 8), the damping value of $\zeta=4.68\%$ also compares favourably with these two guides ($\zeta=4.5$ and 5.0% respectively).

Floor Response

In the SCI and AISC/CISC design guides[25, 26] it is assumed that resonance will occur when one of the harmonic components from the walking frequency coincides with the fundamental frequency of the floor. However, when the fundamental frequency is sufficiently high in comparison with the activity frequency, the floor will exhibit a transient response. In the experimental investigation presented, a number of walking tests were carried out on the eight floors to establish the worst design case (Table 6). In this section, these measurements are compared with the current SCI and AISC/CISC design guides.

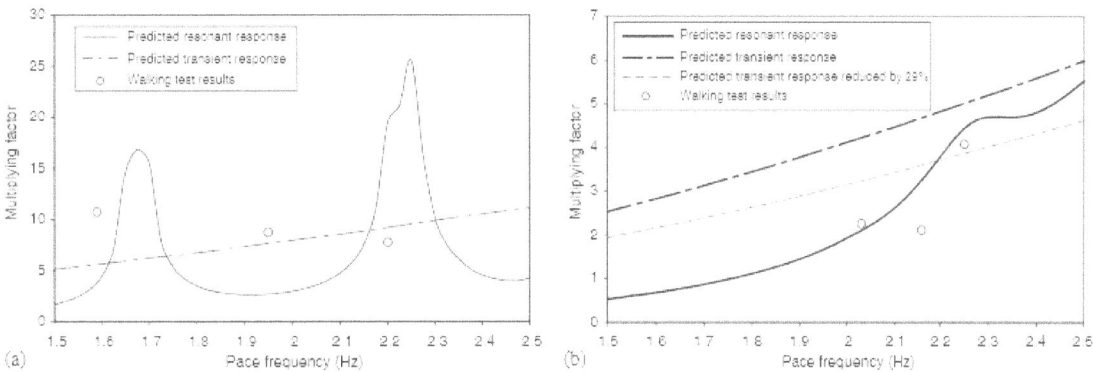

Fig. 11 Predictions for: (a) a floor dominated by a resonant response; (b) a floor dominated by a transient response

In this comparison, the floor response is calculated from the predicted frequency (Fig. 9), forcing function and modal mass. However, to minimize the scatter in the predictions, the measured damping value has been inserted within the respective design equation when appropriate. Since both the SCI and AISC/CISC guide refer to the *z*-axis base curve shown in Fig. 3(d) as a measure of acceptable performance, the values of the predicted floor response have been converted to a common format of multiplying factors (or in the SCI terminology, 'response factors') in Fig. 10. As can be seen from Fig. 10, for the eight floors that were subjected to walking tests, the design methodology given in both the SCI and AISC/CISC guides produce conservative predictions in the majority of cases.

By comparing the measured and predicted values, the average model factor for the SCI predictions is 0.67, which implies that the SCI method is on the conservative side. However, as can be seen from Fig. 10(a), the scatter between the predicted and measured multiplying factors is vast, and corresponds to a coefficient of variation (COV) of 83%. While, when comparing the predictions of floor response using the AISC/CISC methodology, the average model factor is 15.91, which indicates that the method is unconservative. Also, the corresponding COV value of 270% shows that the scatter between the experimental and predicted response factors is immense. However, by considering Fig. 10(b), the data point that has the most significant influence on the statistical properties is from the floor with fundamental frequency above 9.0–10.0 Hz (Floor 7), where the response was dominated by impulsive excitation. By ignoring this particular result, the average model factor reduces considerably to 0.72 and a COV of 99%, which is broadly in line with the predictions given by the SCI guide.

Comparison of test results with numerical analysis techniques

Finite element methods of analysis are now available in most engineering software packages, and are becoming more widespread in use for practical engineering structures. Use of finite element models for considering floor vibrations offers opportunities to the designer, by allowing a more realistic consideration of the floor structure than can be achieved with simple hand methods of analysis. Young[49] has developed a design methodology that is based on the results from modal analyses, which has recently been shown to produce very good agreement with measurements on a laboratory floor[44].

To investigate the highest level of accuracy that may be expected to be achieved with predictive techniques, the eight floors that were described in the previous section were analysed using the methodology described by Young[49]. The finite element modelling followed the general approach that has been described in other investigations[44, 50]. Assumptions that are of particular relevance to the modelling of the eight steel–concrete composite floors reported here are as follows:

- The dynamic Young's modulus of concrete was taken to be 38 and 22 kN/mm^2 for normal and lightweight concrete, respectively25.

- Columns were included within the models, which were pinned at their theoretical inflexion points, located mid-height between the floors.

- Owing to the small displacements associated with walking activities, full moment continuity was assumed to exist between the beam-to-beam and beam-to-column connections.

- Façades were considered to offer vertical restraint.

For the calculation of the floor response, the following assumptions were made:

- Because most existing design guidance refers to the work of Rainer et al.27, the resonant response predictions were based on the first four Fourier coefficients reported by these authors (Fig. 5).

- Based on the excellent correlation with measurements on a 'high frequency' laboratory floor 44, the impulsive force model developed by Young49 was used in the transient response predictions.

- To minimize the amount of scatter, as well as to allow true comparisons to be made with the predictions using the current design guides25, 26, the measured damping values shown in Table 5 were used for each of the floors

A typical comparison of the predicted and measured multiplying factors from walking tests is shown in Fig. 11. As can be seen, there is reasonable agreement with the measurements. For Fig. 11(a), where the floor response was dominated by resonant excitation, it can be seen that, had the walking test been conducted at a pace frequency of around 2.25 Hz, a much higher multiplying factor might have been achieved. While for the floor considered in Fig. 11par;b), where the response was transient in nature, the predictions are slightly higher than the measured results. However, this can be attributed to the fact that the impulse force is a design value, based on a 25% probability of exceedance50. As can be seen from Fig. 11(b), by using the average value for the impulse force (i.e. the design impulse force reduced by 29%), the correlation is very good. Furthermore, it is interesting to note that, had a walking test been conducted at the highest pace frequency possible of around 2.4 Hz, a slightly higher multiplying factor might have been achieved.

A summary of the results for the eight floors considered in the previous section is given in Fig. 12; the predicted multiplying factor given in this figure corresponds to the exact pace frequency in the tests that produced the highest measured value. As can be seen from Fig. 12 the predictions using the methodology proposed by Young49 produce very good correlation with the measured values. By comparing the measured with the predicted values, a slightly conservative model factor of 0.98 is achieved with a corresponding coefficient of variation (COV) value of 57%.

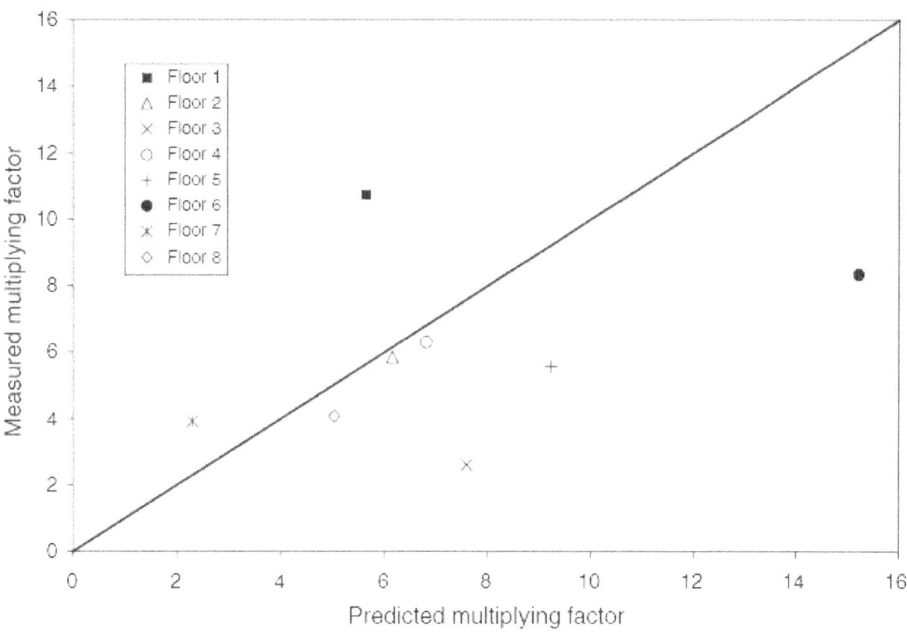

Fig. 12 Comparison of measured floor response with predicted floor response based on finite element models

CHAPTER 5: CONCLUSION

Recommendations and future research

From a consideration of the 17 floors presented here, it is concluded that both the SCI and AISC/CISC approach produce satisfactory predictions of the fundamental frequency for steel–concrete composite floor systems. Therefore, for cases where specifications or codes of practice51, 52 state frequency limits for floors used in certain applications (e.g. dance-floors and aerobics areas subjected to synchronized crowd movement), both the SCI and AISC/CISC guide may be used with confidence.

With respect to damping values, it would appear that the values given in the SCI and AISC/CISC are appropriate for composite floors, and are therefore recommended for design. Also, it would appear that the presence of false flooring, ceiling and services do not contribute significantly to the damping characteristics of the floor. Unfortunately, although the majority of floors considered in this paper were tested in their bare state, it is useful to note that the lowest design damping value that may be considered in steel–concrete composite floors is around 1.0%, which gives an upper bound to the dynamic magnification factor for cases when the floor is excited to resonance. It may also be useful for designers to consider this bare condition, as adverse comments could be raised over the acceptability of the floor, before the building is completely fitted out.

Both the SCI and AISC/CISC predictions of the floor response produced a wide scatter when compared with the measured results presented within this paper with a COV of 83 and 270%, respectively. However, on average, this scatter was compensated by the fact that the predictions were very conservative. As changing the magnitude of response by a factor of two is equivalent to only a marginal change in human reaction25, this level of conservatism may not be a concern for office floors. However, for sensitive environments such as residential dwellings or hospitals the requirement for very low responses may cause the floor design based on these guides to be structurally inefficient, owing to the need for a high level of mass.

Predictions based on finite element analyses appear to offer the most efficient way of designing floors for vibrations, and are of particular use in considering floors with closely spaced modes of vibration. As a consequence of this, response predictions based on analyses of this type are recommended. For the eight floors considered in this paper, such an approach led to very good predictions when compared with the measured results, with an average model factor of 0.98 and a COV of 57%.

In the comparisons made in this paper, the worst design case (i.e. highest multiplying factor) was taken as the measure of the floor performance. However, in reality, it is unlikely that pedestrians will always pace at a frequency that will produce the highest response. As a consequence of this, probabilistic techniques could be considered in future work to produce more realistic estimates of the floor response in service.

In future work, analyses in the time domain may produce much more realistic predictions of the floor response, and could be used with greater confidence in assessments using vibration dose values. However, such an approach may be very expensive and time consuming, and would appear to be unnecessary for practical design.

Conclusions

Vibrations in steel–concrete composite floors are not a new phenomenon. Indeed, the first guidelines for considering this serviceability limit state were developed over 30 years ago.

However, as slender floors are increasingly being constructed, it is essential that a serviceability assessment of the floor vibration from walking activities be fully considered at the design stage.

By comparing the predictions from current design guides with the results of an extensive programme of *in situ* testing, it has been shown that the current guidance is very conservative. This is primarily due to the fact that it is difficult to develop simple methods for assessing the response to walking activities because of the complexity of the behaviour that exists in most modern floor layouts.

The author is currently engaged in a European project, which will hopefully address some of the weaknesses that exist in the current design methods. However, it is likely that hand methods of analysis will only be really appropriate for floors that possess regular grids. Therefore, in future, it may be necessary to use a more sophisticated method based on finite element analysis as part of the design process.

REFERENCES

Dallard P, Fitzpatrick AJ, Flint A, Le Bourva S, Low A, Ridsdill Smith RM & Willford M. The London Millennium Footbridge. The Structural Engineer 2001: 79 (22): 17-33.

Canadian Standard CAN3-S16.1-M89. Steel structures for building (limit states design) Appendix G: Guide for floor vibrations, 1989 Canadian Standards Association, Toronto.

Johnson RP. Composite Structures of Steel, Concrete Volume 1: Beams, Columns, Frames and Applications in Buildings. London. Constrado Monographs, 1975.

Patrick M & Poon SL. Composite Beam Design and Safe Load Tables. New South Wales: Australian Institute of Steel Construction, 1989.

ENV 1993-1-1. Eurocode 3 Design of steel structures: Part 1.1: General rules and rules for buildings 1992, European Committee for Standardization, Brussels.

prENV 1993-1-1. Eurocode 3 Design of steel structures: Part 1.1: General rules and rules for buildings, 2003. European Committee for Standardization, Brussels.

Reiher H & Meister FJ. The sensitiveness of the human body to vibrations (Empfindlichkeit des menschen gegen erschutterung). Forshung (VDI) 1931; 2 (11): 381-386.

Translation Report F-TS-616-RE 1946. Headquarters Air Material Command, Wright Field, Dayton, Ohio.

Lenzen KH. Vibration of steel joist-concrete slab floors. Engineering Journal (AISC) 1966: 3 (3): 133-136.

Allen DE & Rainer JH. Vibration criteria for long-span floors. Canadian Journal of Civil Engineering 1976: 3 (2): 165-173 Pernica G & Allen DE.

Floor vibration measurements in a shopping centre. Canadian Journal of Civil Engineering 1982: 9 (2): 149-155.

www.ingramcontent.com/pod-product-compliance
Lightning Source LLC
Chambersburg PA
CBHW080847170526
45158CB00009B/2665